NATURAL DISASTERS
MEETING THE CHALLENGE

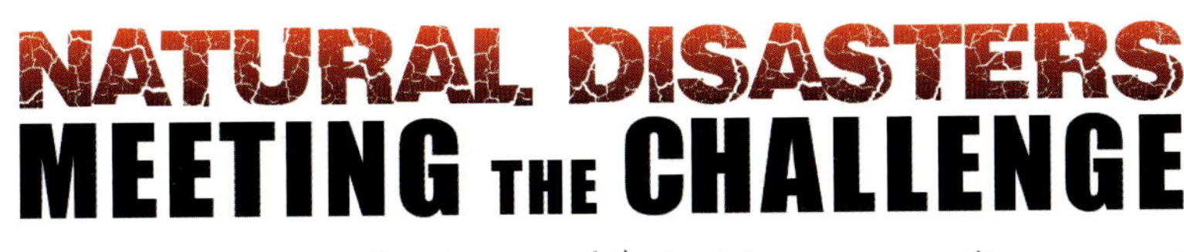

LANDSLIDE AND AVALANCHE READINESS

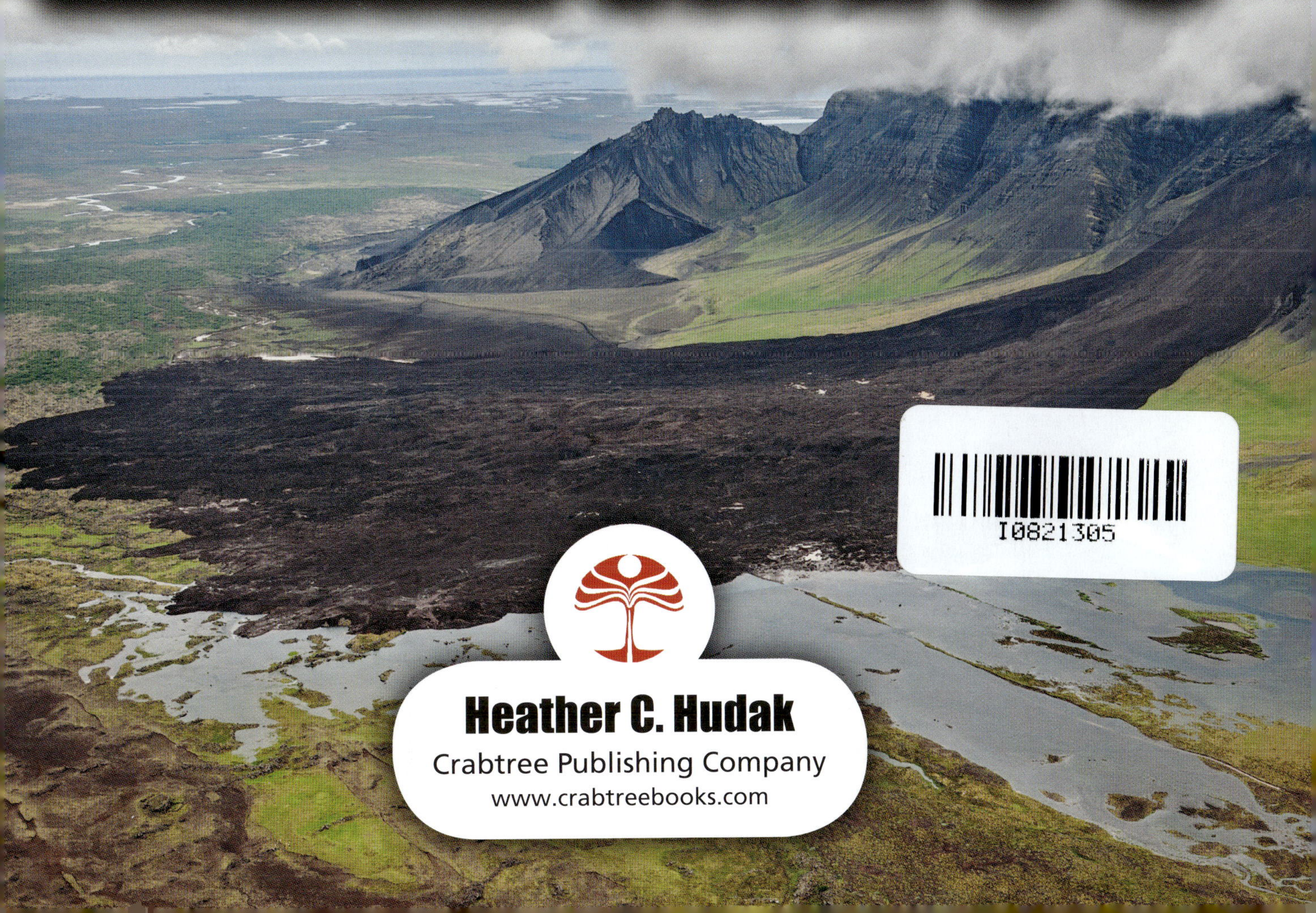

Heather C. Hudak

Crabtree Publishing Company
www.crabtreebooks.com

CRABTREE
PUBLISHING COMPANY
WWW.CRABTREEBOOKS.COM

Author: Heather C. Hudak

Series research and development: Janine Deschenes, Reagan Miller

Editorial director: Kathy Middleton

Editor: Ellen Rodger

Proofreader: Roseann S. Biederman

Design and photo research: Katherine Berti

Images:

Alamy
Foto Arena LTDA/Alamy Live News— Gaston Brito: p. 37
Tjetjep Rustandi: p. 16
Associated Press: The Deseret News, Ravell Call, File: p. 19
Flickr
Mark Goble: p. 22
Getty Images
Frank Bienewald: p. 4
Raul Arboleda: p. 41
STR: p. 7
Ground Truth Trekking—Bjørn: p. 35
InSAR Norway—Screen Shot 2020-01-15 at 9.53.44 AM: p. 26 (top)
Italian Space Agency: p. 26 (bottom)
JOHN L. DENGLER / DENGLERIMAGES.COM: p. 21 (top)
NASA Earth Observatory: p. 27
National Snow and Ice Data Center: Ted Scambos and Rob Bauer: p. 6
Natural Hazards and Earth System Sciences—T. Oppikofer, M. Jaboyedoff, L. Blikra, M.H. Derron, and R. Metzger: p. 29 (bottom)
Shutterstock
dvlcom: p. 39 (bottom)
Thomas Dekiere: p. 5 (bottom)
Wolfgang Simlinger: p. 46
U.S. Geological Survey: p. 28, 36
Ben Pauk: p. 24 (bottom)
LANDSAT: p. 17 (bottom)
L.M. Smith, Waterways Experiment Station, U.S. Army Corps of Engineers: p. 20
Mark Reid: p. 24 (top)
Sound Waves: p. 23
Wikimedia Commons
Asad Amir: p. 31 (bottom)
Frode Inge Helland: P. 29 (top)
Haneburger: p. 13 (bottom)
ilker ender: p. 18 (top)
Nolispanmo: p. 38 (center)
Tankei Inoue: p. 18 (bottom)
Thehunzailad: p. 31
U.S. Coast Guard District 11 PADET Los Angeles District 11: p. 21 (bottom)
www.avalanche.org—Screen Shot 2020-02-18 at 1.47.36 PM: p. 25 (inset)
All other images by Shutterstock

Library and Archives Canada Cataloguing in Publication

Title: Landslide and avalanche readiness / Heather C. Hudak.
Names: Hudak, Heather C., 1975- author.
Description: Series statement: Natural disasters: meeting the challenge | Includes bibliographical references and index.
Identifiers: Canadiana (print) 20190231408 | Canadiana (ebook) 20190231424 | ISBN 9780778774105 (softcover) | ISBN 9780778774068 (hardcover) | ISBN 9781427125071 (HTML)
Subjects: LCSH: Landslides—Juvenile literature. | LCSH: Avalanches—Juvenile literature. | LCSH: Emergency management—Juvenile literature. | LCSH: Hazard mitigation—Juvenile literature.
Classification: LCC QE599.A2 H83 2020 | DDC j363.34/9—dc23

Library of Congress Cataloging-in-Publication Data

Names: Hudak, Heather C., 1975- author.
Title: Landslide and avalanche readiness / Heather C. Hudak.
Other titles: Natural disasters: meeting the challenge.
Description: New York, New York : Crabtree Publishing Company, [2020] | Series: Natural disasters: meeting the challenge | Includes bibliographical references and index.
Identifiers: LCCN 2019052942 (print) | LCCN 2019052943 (ebook) | ISBN 9780778774068 (hardcover) | ISBN 9780778774105 (paperback) | ISBN 9781427125071 (ebook)
Subjects: LCSH: Landslides--Juvenile literature. | Landslides--Prevention--Juvenile literature. | Avalanches--Juvenile literature. | Avalanches--Prevention--Juvenile literature. | Emergency management--Juvenile literature.
Classification: LCC QE599.A2 H84 2020 (print) | LCC QE599.A2 (ebook) | DDC 551.3/07--dc23
LC record available at https://lccn.loc.gov/2019052942
LC ebook record available at https://lccn.loc.gov/2019052943

Crabtree Publishing Company

www.crabtreebooks.com 1-800-387-7650

Printed in the U.S.A./042020/CG20200224

Published in Canada
Crabtree Publishing
616 Welland Ave.
St. Catharines, Ontario
L2M 5V6

Published in the United States
Crabtree Publishing
PMB 59051
350 Fifth Avenue, 59th Floor
New York, New York 10118

Published in the United Kingdom
Crabtree Publishing
Maritime House
Basin Road North, Hove
BN41 1WR

Published in Australia
Crabtree Publishing
Unit 3–5 Currumbin Court
Capalaba
QLD 4157

Contents

Landslides and Avalanches

It was one of the worst natural disasters in the history of modern India. In June 2013, heavy rains flooded rivers and streams in the Himalaya mountains. The waters formed a small lake in the Kedar valley in the country's Uttarakhand state. Soon, these rains led to landslides that carried mud, boulders, and other debris down mountainsides. The landslides breached the lake boundary, and a wall of mud and water then buried the town of Kedarnath, killing 5,700 people.

Landslides, like the one that wiped away Kedarnath, and avalanches, are destructive natural **forces**. They alter the landscape, and cause great harm to people, property, and the environment. In fact, landslides and avalanches kill thousands of people around the world each year, including about 25 to 50 in the United States alone.

Landslides move slower than avalanches at about **31 MILES PER HOUR** (50 kph).

Kedarnath after the slides.

Landslide vs. Avalanche

Both landslides and avalanches are natural **geological** events. They share many similarities, but they are also quite different in other ways. Landslides involve the mass movement of soil, rock, dirt, and debris down a slope. They happen when the forces acting on the slope, such as **gravity**, are stronger than the materials that make up the slope.

An avalanche is a large mass of rock, ice, snow, and other earth materials that flows quickly down the side of a mountain. Avalanches can happen any time of year but are most likely to occur in winter. Common landslide and avalanche triggers include rainfall, snowmelt, earthquakes, volcanic eruptions, and human activities. Any of these triggers can cause the ground to crumble and slide downhill.

Death and Destruction

As landslides and avalanches move downhill, they collect other loose materials along the way. This causes them to grow in size. The high speed, **volume**, weight, and **momentum** of these materials make landslides and avalanches unstoppable. Landslides and avalanches cause billions of dollars in damage each year. People and houses are often buried by rocks, mud, snow, and debris.

*Landslides and avalanches can destroy buildings and knock out **utilities**.*

Roadblocks caused by landslides and avalanches can take days or even weeks to clear and repair damage. If there is no other route, people and goods cannot get in or out of the area. This can lead to shortages of food, fuel, and other supplies people need to survive.

Geological Events

Avalanches and landslides are natural events that have happened for billions of years. They can be small and cause little damage, or they can be massive, causing extreme harm to people and property. At the same time, landslides and avalanches bring benefits to people who live in high-risk areas. Sometimes, natural events are necessary to restore the environment. They create **fertile** land that is ideal for farming, for instance.

Natural events become natural disasters when they have a negative effect on humans. Landslides and avalanches are not always unavoidable. There are things people can do to both trigger and stop landslides. By understanding natural events of the past, we can better predict and prevent natural disasters. This helps reduce damage and loss of life.

CASE STUDY

Up Close and Personal

Much of what we know about avalanche science comes from the knowledge of people who live near mountains, as well as scientists who have made it their life's work studying them. As more people live near mountains, or spend time in mountainous areas skiing and snowmobiling, avalanche science has become more important. But it is also dangerous work. Avalanche researchers and forecasters work "in the field" on mountains and in backcountry ski areas. There, they assess avalanche risk and determine what will help people who live in avalanche-prone areas.

The scientists at the U.S. Forest Service's National Avalanche Center work with university researchers and avalanche forecasters to share information, tools, and techniques for dealing with avalanches and avalanche risk. Sometimes, that means using military weapons to set off controlled avalanches before they happen naturally.

A National Snow and Ice Data Center researcher carves through layers to take a snow sample in an Antarctic snow pit. Studying the properties of snow and ice can help researchers better understand avalanches.

25 to **30** people die in avalanches each year in the U.S. Avalanches kill more people in the country's national forests than any other natural hazard.

Heavy snowfall in the Himalayan mountains of Pakistan led to avalanches that killed 41 people in January 2020. Local residents conducted strike team shovelling searches for victims. This method creates a horizontal ramp to rescue people buried instead of digging straight down.

CHAPTER 1

The Science Behind Landslides and Avalanches

Landslides and avalanches can be extremely dangerous. They often happen quickly and without warning. When debris or snow slides down a slope, they can reach speeds of 124 to 186 miles per hour (200 to 300 kph).

Mass Movements

There are many forces that hold rock, dirt, and clay together. **Friction** is one of them. It helps keep mud and soil in place. Sometimes a weak piece of rock, soil, or dirt breaks away from the more stable ground beneath it. Gravity causes it to pull away from the land surface. It then forms a huge mass as it slides downward and outward.

Landslide is a general term used to describe any kind of mass movement of earth. Avalanches are mass movements of snow, ice, and rock. Mass movements mostly take place on mountains or hills. They can also happen in low-lying areas where there is a strong slope, such as dig sites for roadways, buildings, or river banks.

SCIENCE BIO

Simon Trautman, Avalanche Specialist

Simon Trautman has a mission: to educate people—especially people who ski, snowboard, and spend time in snowy, mountainous areas—about avalanche hazards, forecasting, and safety. An avalanche specialist and director of field operations and forecasting at the U.S. Forest Service's National Avalanche Center, Simon studies avalanches. He writes about them and speaks about them at avalanche workshop and safety events. Simon's message is clear: "if you have mountains and have snow, you'll have avalanches." The key to surviving them may be knowledge and safety planning. A geologist by training, Simon studied wet avalanches and began work as an avalanche forecaster for the Moonlight Basin Ski Patrol, the Colorado Avalanche Information Center, and the Sawtooth Avalanche Center. He believes that avalanche risks can be greatly reduced and lives saved by informing and educating people about the avalanche hazards.

Causes of Landslides

There are a few key causes for a landslide. Water is one of the most common. Heavy rains or snow and ice melt can cause earth to become heavy. This makes it more likely to break away from the surface. **Seismic activity**, such as earthquakes, are another way landslides start. They cause **vibrations** in the ground that reduce the friction holding materials to the surface. Wildfires destroy plants and trees that root into the earth and hold the soil in place. Without vegetation, the land is more likely to loosen and crumble. The explosive force of volcanic eruptions can also cause loose materials to break away.

Types of Landslides

Landslides are classified by the kinds of materials they contain. This includes either rock, earth, or debris. They are also grouped by how they move.

Falling and Toppling Rock

Falling happens when a piece of rock or soil suddenly breaks away from a steep surface, such as a cliff. It then falls to the ground. It may bounce or roll when it hits the surface below. It may even break apart and continue to move. Toppling takes place when a piece of rock or other earth material rotates forward from a land surface and falls downward. This is similar to the way a domino falls.

Spreading Earth

Spreading involves the movement of earth material sideways from the surface. It happens when a strong force, such as an earthquake, makes the ground move quickly. Spreading mostly takes place on flat land or gently sloping areas. Flowing happens when earth materials mix with water and other liquids causing them to move quickly, or flow, down a slope.

Sliding

Sliding occurs when a piece of surface material separates from the ground below and slides away from it. There are two types of sliding. Rotational slides move downward and outward over a spoon-shaped surface that curves upward. Translation slides move downward and outward along a fairly flat, sloping surface.

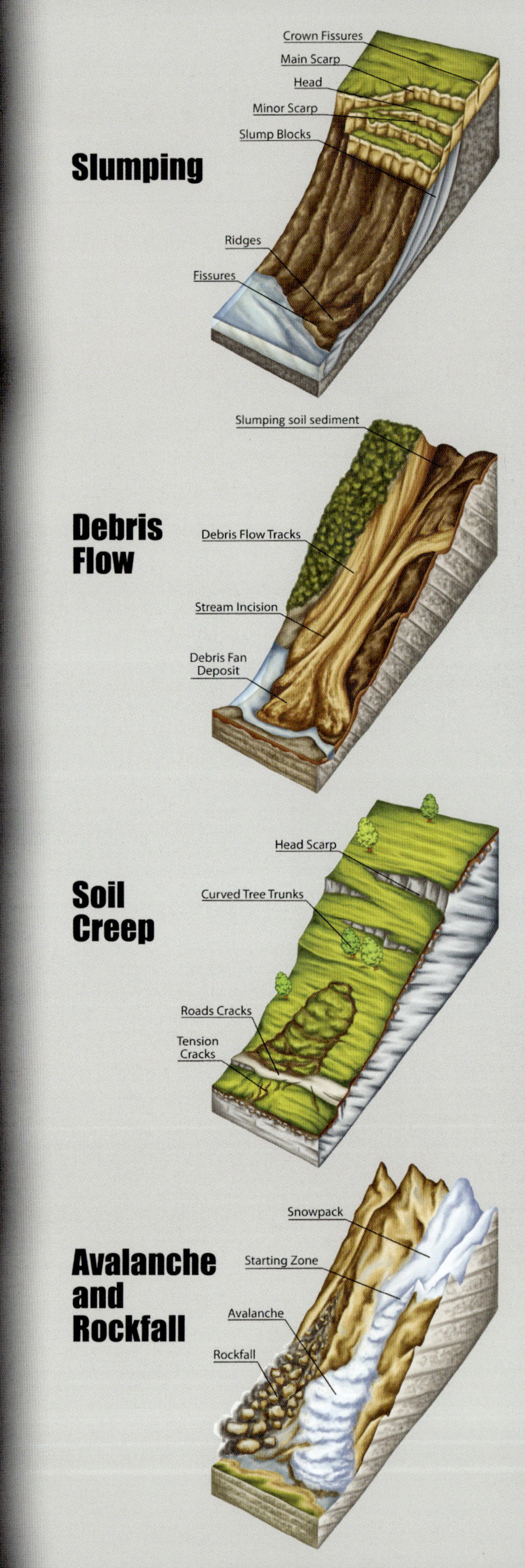

CASE STUDY
The Hope Slides

At about 4 a.m. on January 9, 1965, a small landslide occurred on the highway near Hope, British Columbia, Canada. While crews were clearing the debris away for vehicles to pass through, a second slide took place about three hours later. The massive slide hit Outram Lake with such a powerful force that water shot across the valley. It wiped out plants and trees. Four people were killed, two of whom were never found.

The Hope landslide was the largest landslide recorded in Canada. It was caused by unstable ground.

For years, scientists believed the slides were caused by **tremors**. Seismographs had recorded some in the area at the same time as the slides took place. Recent studies show the vibrations were caused by the slide itself. The weather had been colder than normal for weeks prior to the slide. Scientists today think water may have seeped into **faults** in the slope of Johnson Ridge and froze there. The added pressure helped trigger the slide. The ground on the mountain was already unstable due to years of **weathering** and **erosion**. At first, a new road was built through the slide site. Over time, the route has been changed to bypass the site.

55 MILLION TONS

(50 million metric tons) of rock, dirt, and debris slid down the side of a mountain during the Hope Slides.

Avalanche Ingredients

There are three main ingredients needed for an avalanche to take place. The first ingredient is a large buildup of snow. As fresh snow falls, it piles on top of other layers of snow to create snowpack. Some of the layers are stronger than others. They may contain everything from soft powder to hard ice crystals.

The second ingredient is a weak layer of snow. This layer may be made up of hoarfrost, icy patches, or softer snowpack that is not as strong as the layers on top of it. The weak layer is buried under the snowpack and cannot be seen from the surface. Finally, you need a surface with a slope of 30 degrees or more.

Mix it Up

When all of these ingredients are in place, the conditions are right for an avalanche. All that is needed is something to trigger the slide by putting pressure on the weak layer. It is possible for a rock or a piece of ice to roll onto the weak spot and set the avalanche in motion. The snow may also start sliding on its own for no reason. However, these are unlikely triggers. Most avalanches are caused by human activities, such as skiing or snowmobiling.

Only **3 PERCENT** of avalanches take place on slopes that are **LESS THAN 30 DEGREES**.

Hoarfrost is frozen water vapor.

Types of Avalanches

There are three main types of avalanches. Powder avalanches are most common after a heavy snowfall on a smooth, icy surface. The snow cannot grip the surface beneath it and begins to give out under its own weight. Powder avalanches start in one place and collect snow as they slide down the slope. This is known as the snowball effect because the avalanche grows larger in size as it moves. Powder avalanches can reach speeds of 62 to 186 miles per hour (100 to 299 kph). They look like a cloud of fluffy powder but are very violent underneath.

Slab avalanches most often happen after a fresh snowfall. A piece of the slab breaks aways. It moves down the slope as one big chunk or it breaks into smaller pieces as it moves. Most slabs are about half the size of a standard football field. Slab avalanches are most common in the winter. They can reach a speed of 20 miles per hour (32 kph) within the first three seconds. After about six seconds, their speed increases to about 80 miles per hour (128 kph). When the slab reaches 220 miles per hour (352 kph), it appears to shatter like glass.

Wet avalanches happen when temperatures rise, such as during the spring thaw or when bright sunlight and heavy rains cause water to seep through the snow. As the snow melts, it gets heavier and begins to slide. Wet avalanches tend to move slowly due to the heavy weight of the wet snow. They travel at speeds of 10 to 20 miles per hour (15 to 30 kph).

Avalanche Features

There are three main features of an avalanche. The starting zone is the place where the avalanche begins. It is the weak point where the snow breaks away. Most times, the starting zone is located high on a slope above the treeline and near a cliff edge. However, the starting point can be any place on the slope if the conditions are right.

The path the avalanche follows down the slope is known as the track. Researchers look for clearings and downed trees to provide clues about the avalanche track. The run out is the place where the avalanche comes to a stop. It is easy to locate by the snow, debris, and dirt that piles up when the avalanche stops sliding.

300,000 CUBIC YARDS (230,000 cubic m) of snow can be released in a large avalanche in North America.

A massive mudflow pushed through a mountain stream in the Austrian Alps.

Other Mass Movements

Landslides and avalanches are not the only kinds of mass earth movements. Mudflows form on slopes where there are not a lot of plants to hold the soil in place. Heavy rains mix with silt, soil, and rock to create a thick mixture similar to pancake batter. Mudflows move with great force along river channels, streams, ravines, and canyons. They can be deadly and fast and often wash out bridges and other structures.

CASE STUDY

Using GPS on Landslide Hotspots

The San Francisco Bay area is a hotspot for landslides. Scientists have been researching the causes and effects of slow-moving landslides in the area for decades. Until recently, they still did not know much about them. New technologies allow scientists to study Bay area landslides in more detail. Scientists use the **Global Positioning System (GPS)** and Interferometric Synthetic Aperture Radar (InSAR) to learn about the forces that act on the land. InSAR is a device that measures radar signals to determine the strength of a landslide. They have found that large amounts of **precipitation** combined with seismic activity have led to a higher rate of landslides in the area.

Slumps and creeps are slow-moving masses of earth down a slope. They are found in places where there is a large amount of erosion or weathering. Slumps move as one large block. They often take place when a portion of the slope is cut out and no support is in place for the remaining land. Some slumps move just inches each year. Others move several feet in a week. Creeps are caused by changes in the environment, such as temperature or soil structure. A glacier is an example of a creep. Creeps move so slowly they are barely noticeable to the unaided eye.

CHAPTER 2

Studying Disasters and Their Effects

Landslides and avalanches are often triggered by other natural events, such as wildfires, earthquakes, volcanoes, floods, hurricanes, and **tsunamis**. As a result, they do not get as much attention in the news as the original event even though they may cause more death and destruction.

Scientists study past landslides and avalanches in order to help them understand how, when, and why they happen. The more they know about these natural events, the more they can do to help prevent natural disasters.

Volcanic Activity

A volcano is an opening, or **vent**, in Earth's crust that leads to a chamber of hot molten rock, or **magma**, deep below the surface of the land. As magma rises to the surface, it gets injected into cracks on the slopes of the volcano. This can weaken the structure of the volcano, causing a landslide or avalanche. When volcanoes erupt, hot ash, lava, rocks, and other debris spew from the vent. Sometimes, the top of the volcano even caves in or collapses. Other times, a few rocks simply fall from the rim of the crater. Any of these events can trigger a mass earth movement.

An eruption and summit collapse of the Anak Krakatau volcano in Indonesia in December 2018 triggered a landslide that led to a tsunami that killed more than 400 people. It reduced the summit from 1,108 feet (338 m) to 360 feet (110 m).

Underwater landslide-triggered tsunami waves can reach heights of **100 FEET** (30.5 m) and travel at speeds of **500 MILES PER HOUR** (805 kph).

Lahar Mudflows

Lahars are a type of mudflow that form on volcanoes. They are made up of lava, mud, rock, and other volcanic debris that mixes with water from rain, snow, or melting glaciers. Lahars move as fast as 20 to 40 miles per hour (32 to 64 kph). They are very powerful since they contain a lot of heavy rocks and mud. Lahars can move boulders, destroy houses, and tear trees from the ground. Pyroclastic flows are another type of mass earth movement that happens on volcanoes. They contain hot ash, gases, and lava that flow down the slopes at speeds of up to 450 miles per hour (724 kph).

Kilauea Volcano

Kilauea Volcano in the Hawaiian Islands is the most active volcano in the world. It erupted nonstop from 1983 to 2018. Kilauea is home to a landslide known as the Hilina Slump. There are many large faults on the south side of the volcano. Several cliffs sit atop the faults. Movement along the faults caused by earthquakes has caused the slump to slide closer to the ocean. In 1975, a powerful earthquake moved the slump about 11 feet (3.4 m).

If the slump were to become a fast-moving landslide, the result would be disastrous. It would trigger a massive earthquake and tsunami waves rising 1,000 feet (305 m) high. A similar slide took place about 110,000 years ago. All of Hawaii was submerged by the waves.

Ground Movements

Seismic activity is a common cause of mass earth movements. When rocks suddenly break, or explode, they release energy in the form of seismic waves. These waves vibrate and create earthquakes as they move through Earth's crust. About 50,000 earthquakes take place each year that can be detected without the use of technology. However, only about 100 create major damage. Much of the damage earthquakes produce comes from landslides and avalanches they trigger.

Earthquake Aftermath

There are many examples of mass earth movements started by earthquakes. The Wenchuan earthquake in China's mountainous Sichuan province led to deadly landslides in May 2008. The intense quake happened after two **tectonic plates** collided along Longmenshan Fault. The quake set off massive landslides. Heavy rains that followed the event triggered even more landslides and caused further movements in some existing landslides.

In 2015, Mount Everest in Nepal was struck by a massive avalanche. On April 25, an earthquake rattled Kathmandu Valley. Minutes later, a sheet of rock and ice flowed down Mount Pumori and swept through a busy Mount Everest base camp. At least 20 people were killed and another 120 injured.

In **1888**, *a volcano on Japan's Mount Bandai forced a peak to collapse and fall outward. It triggered a massive avalanche that sped at*

50 MILES PER HOUR

(80 kph), over an area of about

21 SQUARE MILES

(54 sq km), and killing

460 PEOPLE.

CASE STUDY

Landslide Leads to Earthquake

Sometimes landslides can set off earthquakes, too. In April 2013, one of the largest mining-induced landslides in history took place at Bingham Canyon Mine near Salt Lake City, Utah, in the United States. A first landslide happened when the wall of the mine **collapsed**. A second landslide occurred about 1.5 hours later. Each one lasted 90 seconds. A small earthquake happened beneath the mine about 11 minutes after the second slide. It was caused by the fast shift in weight of the earth.

The landslide, measured at 165 million tons (150 million metric tons) of rock, damaged mining machinery but did not harm any of the 500 people who work at the mine. The mine had expected the landslide and had used sensors and lasers to detect movement. Radar technology also allowed the mining company, Kennecott Utah Copper, to detect small changes of the mine wall. They changed the location of roads and buildings months in advance, and **evacuated** workers the day of the slide.

Kennecott's Bingham Canyon Mine is an open-pit copper mine. Open pit mines are where ores are mined from the surface in benches, or levels, that are cut downward. This results in large quarries and holes. Open-pit mines alter the landscape and rock and soil stability.

Water Damage

Precipitation-soaked slopes are one of the top causes of landslides and avalanches. Anytime there are heavy rains, melting snow, flooding, and changes in water levels, slopes can become very wet. The earth and rock beneath are more likely to start sliding under the weight of the water.

Landslides happen more often in places that have a tropical climate and mountainous regions, such as the Philippines, Central and South America, and southeastern Asia. Here, heavy **monsoon** rains, hurricanes, and other weather events combine with sloping landscapes to create the perfect setting for landslides.

Massive Waves

Tsunamis are most often triggered by earthquakes or volcanic eruptions that take place underwater. However, landslides and avalanches can also set off tsunamis. Massive waves can form when large amounts of earth materials slide into the water. In 1958, an earthquake triggered an avalanche in Lituya Bay, Alaska. A 30-foot-high (10 m) wave ripped through the community, killing two people. Today, frequent rock falls and avalanches at Tidal Inlet in Alaska's Glacier Bay National Park pose a tsunami threat to locals.

Heavy rains caused major mudslides in northern Venezuela in 1999. Nearly 200,000 people were evacuated and as many as 30,000 died. About 60 miles (100 km) of coastline was wiped out by the slides.

A landslide at the Lamplugh Glacier at Glacier Bay National Park in Alaska, June 2016

The 2017 Thomas Wildfire in California led to a series of landslides when heavy rains destabilized soil on hillsides that had lost vegetation. The slide killed 23 people, injured 167 others, and damaged hundreds of homes.

Wildfire Risk

Wildfires greatly increase the risk of landslides by damaging trees, plants, and the soil. The heat of a fire can crack rocks and loosen them from the ground, making them more likely to fall. Stumps that help slow the movement of earth during a landslide are burned during wildfires, too. This results in more water runoff during heavy rains and rapid snowmelt. Runoff carries large rocks, trees, and other debris that can trigger a landslide.

CHAPTER 3

Meeting the Challenge

Mass earth movements, such as landslides and avalanches, can happen anywhere in the world. In fact, they are a **hazard** in all 50 U.S. states, especially in Alaska and Hawaii, coastal and mountain regions in California, Oregon, Washington, and the eastern part of the country. It is important to study landslides and avalanches so that communities know the potential risks they face and how to reduce them. It is often impossible to eliminate the threat. But proper planning can help lessen the impact if a landslide or avalanche occurs.

High-Risk Areas

Any place that has a great deal of erosion is more likely to have a landslide or avalanche. Rivers and waves wear away the land to create banks and cliffs where landslides and avalanches often take place. Places were humans have made changes to the landscape also face a greater risk. Clearing forests causes soil **degradation**. Rerouting waterways around roads and buildings leads to more water in areas than are less able to handle it. Open-pit mining creates steep slopes in the ground. All of these factors can create landslide and avalanche hazards.

Terrain traps are landscape, or terrain, features that make avalanches more serious. They include gullies, creek beds, ditches, and depressions that trap moving snow or cause avalanches to accumulate in small areas.

A temperature of **32 DEGREES FAHRENHEIT** (0 °C), plus sunshine or rain, can destabilize snow and set off avalanches.

CASE STUDY

Landslides Brought by Droughts

Similar to wildfires, **droughts** leave landscapes barren and dry. Sudden rains in drought regions with steep terrain can lead to landslides. An example of this was seen during the Mud Creek slide in the Big Sur region of California in 2017.

Heavy winter and spring rains caused 6 million tons (5 million metric tons) of rock and dirt to slide onto a 0.25-mile (0.4 km) section of the highway. The five years leading up to the landslide had been a time of drought in the area. Scientists had not detected any movement before the Mud Creek slide. However, further research showed that the creek had been meandering downward.

Satellite and airborne InSAR were used to measure the land in the areas of the landslide between 2009 and 2017. Scientists found that Mud Creek had been moving slowly downhill by about 7 inches (18 cm) per year. The rains just triggered the stable landslide to become unstable.

Observing the Site

Scientists are always looking for ways to predict where and when landslides will take place, how fast they will move, and how big they will be. Researchers try to get to the site right after a slide so they can view the conditions that caused it before the area is covered in vegetation or faces erosion.

Field Work

Researchers in the field observe the physical state of the slide site. They look at photographs—including aerial photos—to compare what the site looked like before and after the slide took place. They take photos of the current condition of the site, too. Researchers also measure the length, width, depth, and gradient of the slope to map the conditions. They use tapes, drills, and binoculars to figure exactly where the slide struck. Rock hammers and drills are also used to collect samples of rocks and other earth materials. They perform tests on the samples in a laboratory to find out what they are made of and how strong they are. These details can help researchers better understand the slide process.

This United States Geological Survey (USGS) seismometer "spider" tracks ground shaking and is equipped with a global positioning system (GPS) to track subtle ground movement. It was used to track a landslide in the state of Washington.

SCIENCE BIO

Avalanche Center Cofounder

Karl Birkeland is the director of the U.S. Forest Service National Avalanche Center. In college, Birkeland was a member of the ski patrol and he had to learn to deal with the ever-present threat of avalanches. This sparked his interest in snow science. After earning a master's degree at Montana State University, Birkeland took a job as an avalanche specialist at the Gallatin National Forest in Bozeman, Montana. He later cofounded the National Avalanche Center, an organization that studies and provides information on avalanches and avalanche safety. Birkeland also completed a Ph.D. at Arizona State University.

Birkeland's research was originally on creating models for assessing changes in snow stability. He was the first person to try mapping avalanche conditions in the U.S. over a mountain range in a day. Today, Birkeland helps make sure avalanche scientists and centers across the United States know about the newest avalanche technologies and scientific findings. He also studies past weather patterns to see how they affect avalanche conditions. His goal is to help scientists predict avalanches so they can reduce their impact and death toll.

The National Avalanche Center's National Danger Map reports avalanche danger risks in map form so that people can check regional links and get local avalanche bulletins.

Advanced Technology

Researchers rely on different types of technology to conduct their work. Many scientists use satellite data to help find the risk of landslides and avalanches. The Thematic Mapper is a device found on NASA satellites that measures solar radiation as it bounces off Earth's surface. It produces images of Earth that can be used to identify slide areas. Images from the TM are combined with detailed images of the land taken by InSAR instruments. Together, InSAR and TM images can be used to create high-quality landslide maps. The maps help researchers identify danger zones.

InSAR

Synthetic Aperture Radar Interferometry (InSAR) is a technology researchers use to monitor slope stability in high-risk slide areas. InSAR uses radar images of Earth's surface to detect small changes and **deformations** that take place over time. Scientists compare radar signals from two images taken at the same place at different times. Any differences in the signals mean a shift in the landscape. The images can be used to create slope maps.

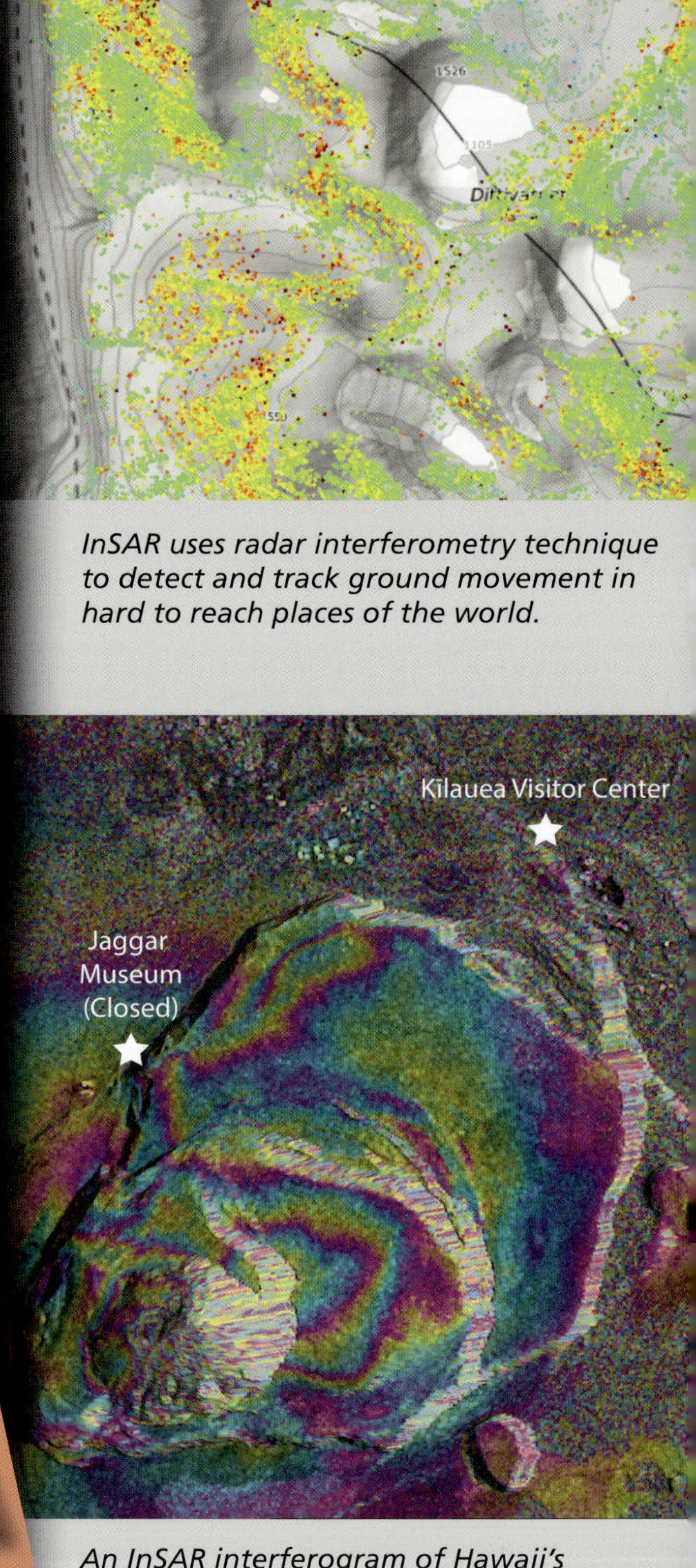

InSAR uses radar interferometry technique to detect and track ground movement in hard to reach places of the world.

An InSAR interferogram of Hawaii's Kilauea Volcano shows ground displacement in different colors.

InSAR can easily detect changes of **0.04 TO 0.08 INCHES** *(1 to 2 mm) and capture new images* **EVERY 2 TO 12 DAYS.**

Fiber-Optics

Researchers use fiber-optic sensors to detect even the smallest shifts in soil. A fiber-optic sensor is a device made up of an **optical fiber** connected to a source of light, such as a laser. It can measure temperature, pressure, strain, vibrations, speed, and more. Researchers embed the sensors in the ground in places where slides are likely to occur. The sensors bend anytime the land shifts. Each movement appears as a loss of light on the sensor. Fiber-optic sensors can be used over large areas to monitor huge slides and small, slow earth movements.

Fiber-optic sensors are smaller, lighter, and more affordable than electrical sensors. They are more durable and can withstand harsh conditions, such as high temperatures. Information gathered by the sensors can be sent to research stations in **real-time** using wireless networks. The data can be used to provide early warning of slides to help save lives.

SCIENCE BIO
USGS

The U.S. government formed the USGS to study the nation's landscapes and help reduce loss of life and property caused by natural events. The National Landslide Hazards Program (LHP) was formed in the 1970s as part of the United States Geological Survey (USGS). The LHP studies landslides and looks for ways to limit their impact. It studies landslide events, performs hazard assessments, and looks for ways to predict and prevents slides from taking place. The LHP also helps educate the public about hazards and provides aid during landslide emergencies.

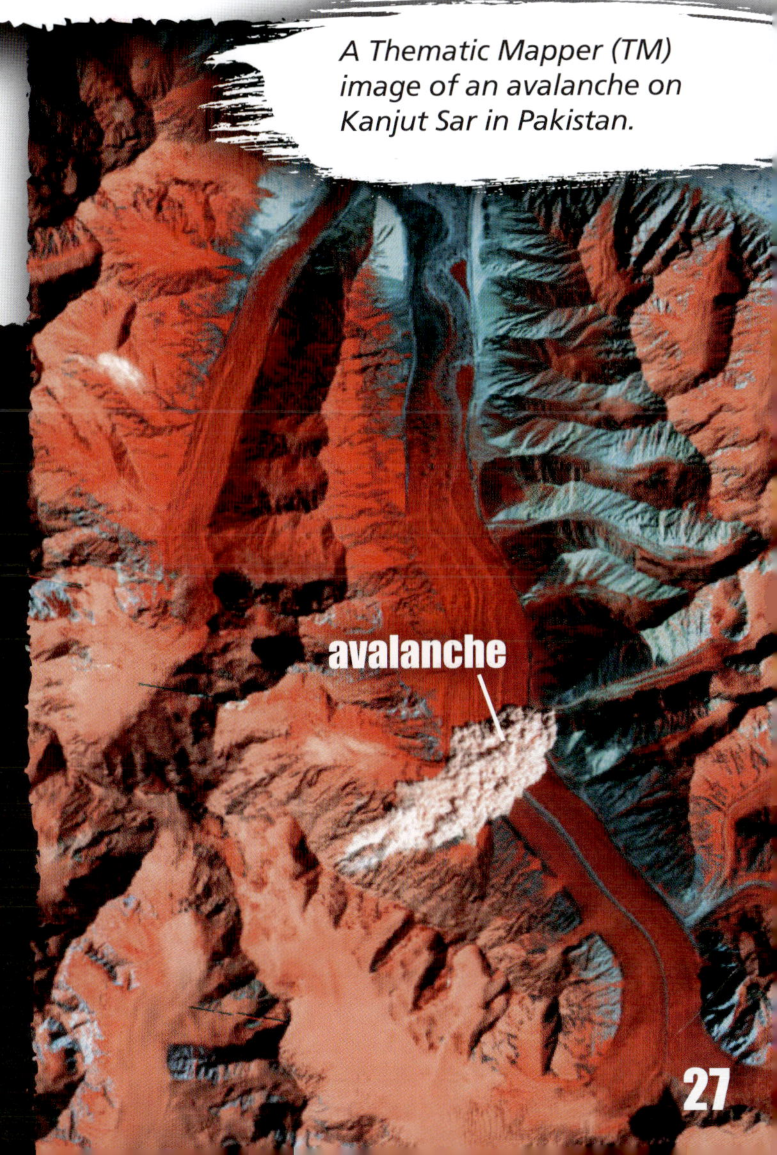

A Thematic Mapper (TM) image of an avalanche on Kanjut Sar in Pakistan.

Landslide Inventory Maps

Landslide inventory maps show the location, date, and types of mass movements that have taken place in an area. Scientists use field surveys, aerial photos, satellite images, historical records, and other information to create a map of slide events. There are many different types of landslide inventory maps. Some include only a small amount of detail. For instance, a simple landslide map might show the locations and dates of past slides. These maps are useful for identifying areas where the slope might be less stable. Other landslide maps contain a lot of information. For example, they might provide data on precipitation levels, slope angle, soil type, type of movement, and seismic activity.

Hazard Assessments

A hazard assessment is a type of inspection. Scientists inspect an area for any slide hazards and the risks they pose. They also identify measures that can be put in place to help protect people, property, and the environment from harm.

Scientists perform hazard assessments to see how likely it is a landslide or avalanche will happen in a certain place. They also try to figure out when the natural event might happen. Hazard assessments are used to help communicate the dangers to locals and find ways to reduce the risk. Hazard assessments are often presented as maps that show debris flows, slide paths, possible rock release points, slope movement, and more.

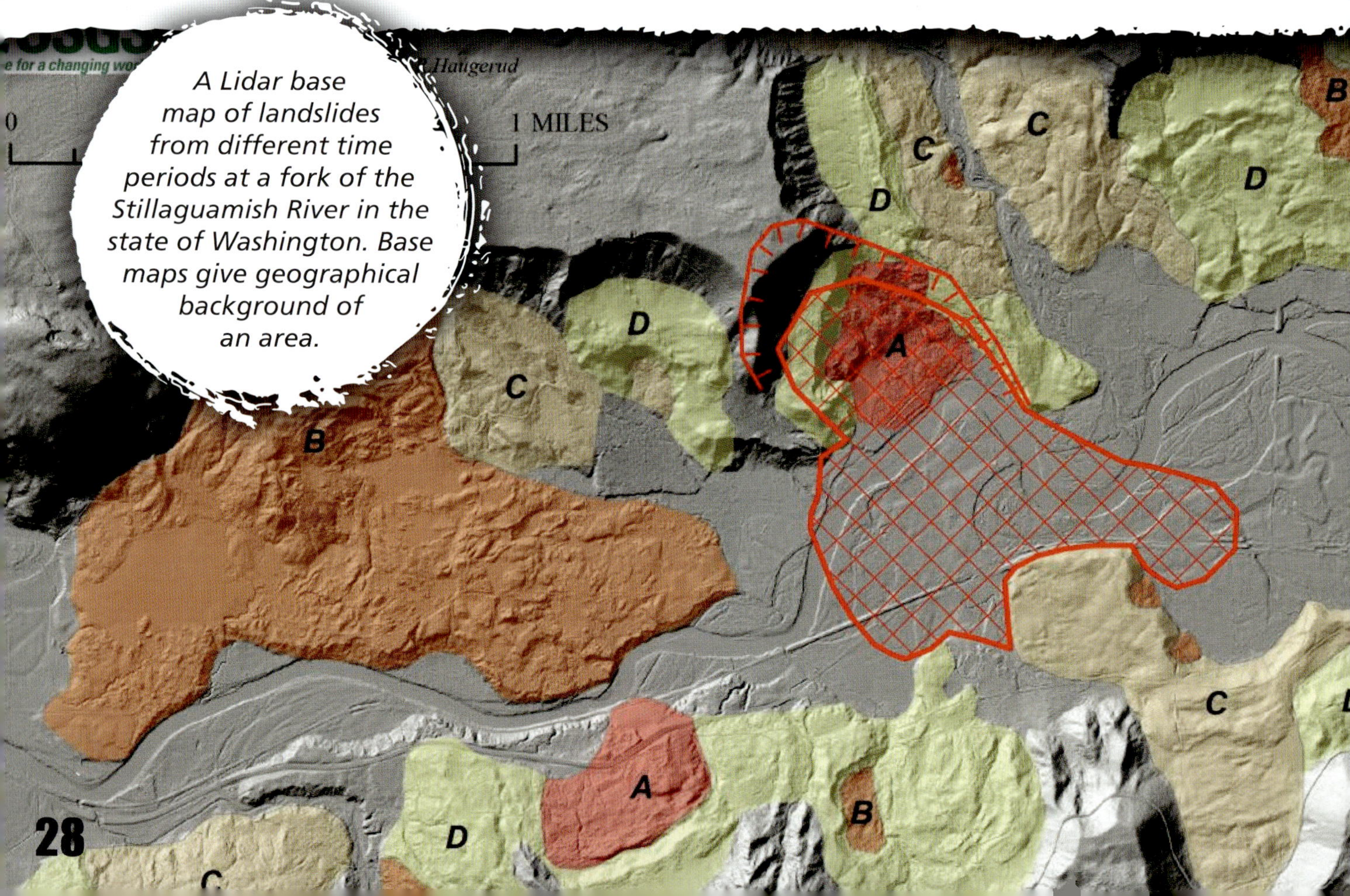

A Lidar base map of landslides from different time periods at a fork of the Stillaguamish River in the state of Washington. Base maps give geographical background of an area.

CASE STUDY

Computer Models of Earth Movements

Geologists at the Vienna University of Technology (TU Vienna) are using computer programs to create models of mass earth movements. The models are helpful for landslide and avalanche hazard and risk assessment. Some of the places the team has modeled include a moving slope above Geiranger Fjord in Norway and the Gschliefgraben, a large landslide system in Traunsee, Austria. The team has adapted existing computer programs to help them **simulate** mass movements. Some slides crash down in one place, while others travel for many miles. The team uses a variety of data to come up with a lifelike model that shows what would happen if a slide took place in the area. Each landslide is very different so each model is unique, too.

Mount Åkerneset, Geiranger Fjord, Norway. The mountain is eroding into the fjord.

Studies show that at least **90 PERCENT** of damage due to landslides **CAN BE PREVENTED** by simply being aware of the hazards and risks.

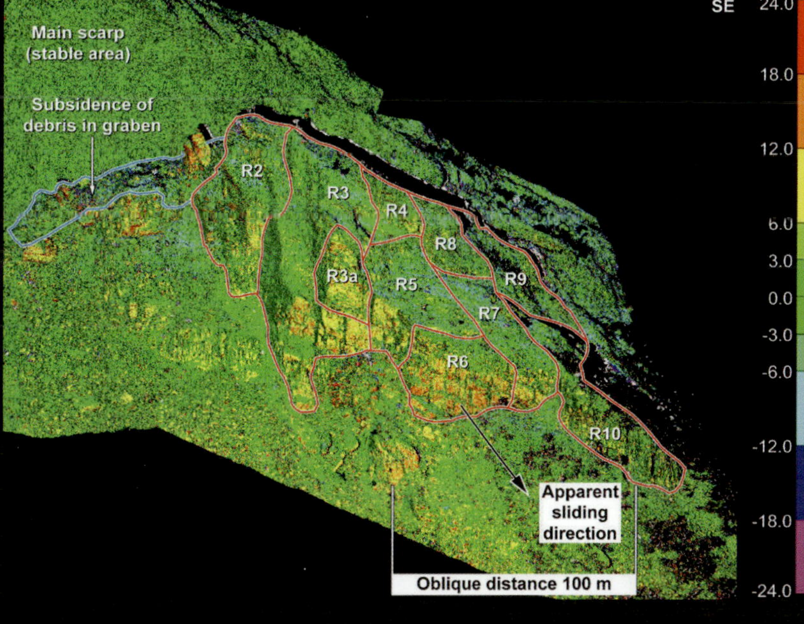

A computer model of Mount Åkerneset helps scientists know what areas are most in danger of landside and how that might affect nearby towns through a landslide-caused tsunami.

Learning From the Past

Areas that have a history of landslides and avalanches are more likely to have future slide events. Understanding the causes and effects of past slides can help prevent future disasters. Researchers use this information to estimate the types, frequency, extent, and consequences of potential slide events. Many of the hazards that could lead to future slides are the result of geologic events that happened long ago. Old slides can also be reactivated.

Researchers compare current landslide and avalanche data to historic slide events to look for patterns. They use landslide inventory maps to find the locations and dates of past events. Newspaper stories are a source of first-hand accounts from the past, that help researchers understand the effects of these events. Researchers also study past natural events, such as volcanic eruptions, earthquakes, floods in the area. Old weather records are also helpful, as they provide the type of data researchers use in modeling, such as rainfall, snow melt, temperature changes, wind speed, and ground freezing and thawing. All these factors contribute to slope instability and the likelihood of future slides.

Disrupting Transportation

Most cities and towns rely on well-worn transportation routes for travel and the import and export of goods. In fact, when building a new community, transportation is one of the top considerations. This includes roads, highways, rivers, and trails. However, transportation routes face major threats from natural events, such as landslides and avalanches.

Remedying Road Collapse

Roads can collapse due to weak and slide-prone soils, blocking important routes. While slide debris can be cleared away in most cases, sometimes there is major damage to the route. It can cost millions of dollars to repair or rebuild. In November 2019, heavy snowfall set off a series of avalanches in Austria and northern Italy. Dozens of major roadways and most smaller routes were completely blocked.

CASE STUDY
Landslides and Highways

GLOBAL WARMING, GLACIAL MELTING, and **DEFORESTATION** *have increased the risk of landslides in recent times.*

In April 2016, 16 people were killed, hundreds injured, and thousands of houses were destroyed by landslides in northern Pakistan. The landslides also caused closures at more than 200 places along the Karakoram Highway—a major travel and trade route through the Karakoram mountains to China. Thousands of people were left stranded. Some roads were cleared and reopened within weeks. However, two parts of the highway were difficult to repair and took months.

Regions accessed by these parts of the highway faced shortages of food and fuel. Wheat and flour were brought in via helicopters and airplanes. In addition, the slides caused damage to power sources and water supplies. In some places, people were without electricity for months. The slides also damaged **irrigation** channels used by local farmers, putting their crops at risk. Landslides have always been a problem in this part of Pakistan.

In 2010, a major landslide destroyed parts of the highway and blocked part of the Hunza River, creating a 13 mile (21 km) lake. The flooding destroyed several villages and farmland and displaced 25,000 people.

Making Plans

As communities grow, space runs out for people to construct new homes, farms, and other buildings. People begin looking for unused land, which may be in high-risk landslide and avalanche zones. In this way, places that were once thought too unsafe become part of the community. The best way to reduce the risks is to avoid them. This means staying out of areas where slides have happened in the past or where scientists have noted a potential threat.

One way governments can help prevent landslides and minimize their impact is by putting land-use policies and regulations in place. Failure to plan could lead to major issues, such as power, water, and sewer outages, and business closures, due to road blockages or damage from a slide.

Landslides and avalanches can happen on stable ground too. Human activities such as watering the lawn, draining reservoirs, and grading slopes can lead to earth movement.

Environmental Issues

People and property are not the only things that suffer during landslides and avalanches. These natural events can have a major impact on the environment as well. Large earth movements have the power to change the face of the landscape. Plants and trees may grow back over time. But some of the changes to mountains, valleys, loss of topsoil, and bodies of water can last forever.

Slope instability in Chernomorsk, Ukraine caused houses to slump and crumble.

Water Pollution

Landslides can fill bodies of water, such as lakes, with large amounts of sediment. The sediment pollutes the water. This reduces water quality. In some cases, rivers may be blocked by debris. It may even form a dam that prevents the water from flowing freely. In most cases, the water will erode the dam over time. In the meantime, fish habitats can be changed. It can also be hard for animals to find another source of drinking water. Sometimes, a lake will form behind the dam. Slumgullion is an enormous slow-moving landslide in the San Juan mountain region of Colorado. It began moving about 700 years ago. The landslide created a dam at the Lake Fork of the Gunnison River to form Lake San Cristobal, the second-largest natural lake in Colorado.

Tearing up Trees

In many parts of the world, slides have a huge impact on forests. They level trees and tear up vegetation. Wildlife lose their main source of food and habitat. They must adapt or move to a new place to survive. Slides are an especially big problem in forests where there is a lot of rainfall and earthquake activity. In 1976, 21 square miles (54 sq km) of forest were wiped out after earthquakes shook Panama.

It can take a single tree as many as **80 YEARS** *to grow after a landslide. An entire forest takes about* **200 YEARS** *to reach maturity.*

CHAPTER 4

Facing Future Disasters

Avalanches and landslides often cannot be prevented. Scientists are always looking for new technologies and methods to help figure out when and where slides will take place so they can reduce the risk to people and property. However, the best way to prevent natural disasters due to landslides and avalanches is to be aware of the hazards and know what to do before, during, and after one occurs.

About **90 PERCENT** of avalanches are due to human activities.

Human Impact

Not all landslides and avalanches are due to natural causes. Changes in the way land is used are a common factor that causes landslides. Digging into snow, using explosives for construction, or taking part in activities such as skiing are known to cause avalanches in high-risk areas. In most cases, victims of the slide accidentally trigger the event.

Four skiers triggered an avalanche in the Blåbærfjellet mountain region of Norway on January 17, 2019. Avalanches are common in the area, taking place at least once per winter. Researchers used data from the sports watch one of the skiers was wearing to recreate their path. The skiers had been walking beneath a cliff when they felt a shift in the terrain above them. The avalanche swept them along for about 0.3 miles (0.5 km) before burying them under 7 feet (2 m) of snow. It was moving at a speed of about 43 miles per hour (70 kph).

A survey team examines the western shore of the Taan Fjord. The team mapped the location where a 2015 landslide occurred. The landslide created a tsunami 633 feet (193 m) high.

Slope failure at the end of Tyndall Glacier sent **180 MILLION TONS** (163 million metric tons) of rock into Taan Fjord, Alaska, on October 17, 2015.

Climate Change

Landslides and avalanches are happening more often and are more powerful than ever before. One of the main reasons is climate change. Global warming is the long-term rise in Earth's average temperature. As a result, there has been a rise in extreme weather events, including more floods, heavier rainfalls, and stronger hurricanes. Warmer temperatures can weaken snowpack. More weak spots mean more opportunities for an avalanche. The rapid thawing of **permafrost** and melting glaciers creates wetter, more unstable slopes as well.

Living in a Danger Zone

People who live in a high-risk zone should always be prepared for a slide. To begin, learn about the geology of the area. Keep up to date on any potential hazards and the likelihood of a slide. It is important not to undercut a steep bank, build near the top or foot of steep slopes, or do anything to increase the flow of water on a slope. All of these actions reduce slope stability. It is a good idea to always have an emergency kit on hand that includes blankets, battery-powered flashlights and radios, bottled water, canned foods, and a first-aid kit. Have an escape plan in place and know who to contact for help if needed.

Warning Signs

Avalanches and landslides are very hard to predict. But there are signs one is about to take place or is already in motion. Look for water or soaked ground in places that are normally dry or a sudden increase or decrease in the water levels of rivers and creeks. Cracks, bulges, or sinkholes in roads and sidewalks and leaning trees, fences, and telephone poles are other signs that the ground is shifting.

At home, watch for cracks in the ground or soil moving away from the base of the house. Doors and windows that are hard to open or close, broken pipes and utility lines, and tilting decks may be signs of a ground movement. Abnormal sounds, such as trees cracking or falling and large boulders knocking together, could also mean the ground is moving. A slide in motion will make a quiet rumble that gets louder as the slide gets nearer.

Be Informed

Many places issue alerts when there is a risk of a landslide or avalanche. These alerts are broadcast on TV and radio news programs and on social media. In the United States, a landslide advisory is sent out when there is the potential for a landslide in a certain region. Often, advisories are issued if heavy rains are expected that could trigger a slide. The statement may include details about what to expect in the event of heavy rainfall and how to stay safe.

A landslide watch means that a landslide could possibly happen. People should be aware of the risk and prepare for it just in case. They should stay up to date on the latest details and check the weather often as any changes could set off a slide. A landslide warning is issued when there is a landslide in action. People should avoid the area and practice extreme caution.

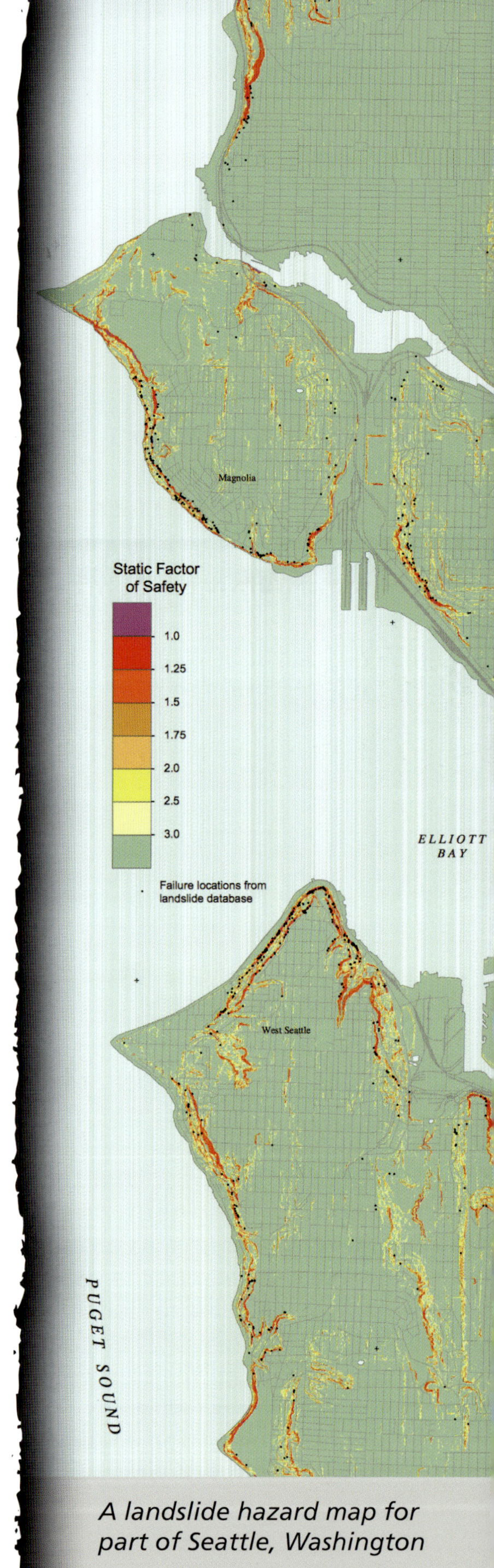

A landslide hazard map for part of Seattle, Washington

CASE STUDY

Using InSAR in La Paz

La Paz, Bolivia, is the highest capital city in the world. It sits amid the mountains at an altitude of 12,000 feet (3,600 m). Many communities are built on steep, unstable slopes. Landslides often sweep through La Paz, especially during times of heavy rains and flooding. They have destroyed countless homes and wiped out important **infrastructure**. The city allows people to build in areas where slopes are known to fail. In some cases, there have been many landslides in the same area. Yet people continue to build there. Scientists have used InSAR technology to study slide movement and slope instability in high-risk parts of La Paz. The results are being used to help city planners manage the land better and find ways to mitigate slide risks. They include creating hazard maps that show the most landslide vulnerable areas and better evacuation planning to help people who live in areas where risks are especially high.

Safety Precautions

People often enter or live in high-risk slides zones on purpose. Some live in risky areas because they have nowhere else to live or because they believe the risk is low or worth taking. Sometimes people use roads and trails that pass through mountainous regions. Other times they hike or ski in areas of high risk. No matter the reason, it is important to take safety precautions when entering a landslide- or avalanche-vulnerable area.

First, it is important to find out if there are any known slide risks in the area that should be avoided. This information can be used to come up with an emergency plan just in case a slide happens. It is also a good idea to tell someone back home about any travel plans so they know where to look if a slide takes place. Be sure to also check the weather forecast for sudden temperature changes or heavy rains that could trigger an avalanche or landslide.

Living in Risk

People in high-risk zones should constantly inspect their surroundings for any loose rocks, cracking ice, or other warning signs. They should always travel with others who can help look for hazards. Carrying special gear is necessary when traveling through high-risk slide zones. A GPS can help a person find their way if they get lost during a slide. A beacon is useful to signal for help, and a shovel can be used to dig through slide debris.

A digital avalanche beacon can help locate people buried under avalanches.

Slope rockfall protection mesh and landslide protection cover the side of a landslide-prone area in Telpaneca, Madriz, Nicaragua.

Emergency Response

The Red Cross and other aid organizations offer relief and assistance to people impacted by natural disasters, such as landslides and avalanches. They provide temporary shelters for people who cannot return to their home due to damage, blocked roadways, or downed utilities. They also bring food and water to people in need and find ways to get supplies into places where transportation routes have been cut off. Aid organizations perform search and rescue missions to help find people who have been buried in rocks, snow, and other slide debris. They also provide medical care to people who have been injured.

An avalanche rescuer uses a probe to locate people in an avalanche simulation exercise.

Stopping the Flow

Landslides and avalanches are powerful forces of destruction. Once in motion they are hard to stop. In some cases, it is possible to stop or reduce the flow of snow, earth, and debris down the slope. Fences, nets, and windbreaks can be used to help prevent earth and snow from building up in certain areas. They also stop it from sliding down a slope or reduce its impact if it does.

Houses sit below a mountain slope with avalanche fences in Hammerfest, Finnmark, Norway.

Setting Off a Slide

One way slides are managed is by triggering small, controlled avalanches at specific times and places. Controlled avalanches are deliberately triggered to reduce the risk or hazard of uncontrolled, or "natural" avalanches. They are set off in snowy, avalanche prone areas by scientists, conservation management officials and ski hill workers. They dig small pits into the snowpack to look for weak layers. They can also use radar to analyze the area. Then they clear the slopes and ensure there is no one in the way of the avalanche path.

Once everyone is safe, they set off explosives to disturb the ground and cause an avalanche. Sometimes, they use helicopters to drop the explosives. Other times, they connect the explosive to a wireless remote so they can set it off from a distance. On small slopes, they may send a person to ski along weak areas instead. The skier tries to stay out of the way of the avalanche. A second skier is on site to help in case of an emergency.

The Gazlex avalanche control system uses a propane and oxygen gas mixture to set off explosions above the snowpack. This releases controlled avalanches, making it safer for skiers and people living in the area.

Planting Trees

Logging and **deforestation** put areas at greater risk of landslides and avalanches. Trees and forests play an important role in preventing landslides by keeping soil and rocks rooted. They absorb water into their roots to help keep the slopes dry. Trees also help reduce the impact of landslides by breaking the flow of moving earth. Planting new trees in places where they have been cut down can help prevent landslides. Trees and shrubs can also help reduce the amount of sediment that pollutes rivers and lakes after a landslide takes place.

CASE STUDY
Landslide Resilience Projects

Landslides cause enormous damage and destruction across the South American country of Colombia each year. Nearly 45,000 homes in Medellín alone are at high risk. Informal settlements, or housing built without approval or that doesn't follow building codes, are most at risk. In Medellín, many people live in makeshift housing on steep, unstable slopes. Landslides are an extreme hazard—so much so that locals began asking the government for help to mitigate the risks.

The Global Challenges Research Fund is a U.K.-based program that helped start a "Resilience or resistance" project to develop plans to help reduce landslide risk in Medellín. Community leaders, local residents, and experts from across Colombia and the U.K. worked together to come up with ideas to help reduce the risk of landslides in informal settlements. They started by mapping out areas where water control was a problem and the ground was cracking or breaking. They set up a community-based team of researchers who took photos and videos of changes in the landscape and sent them to experts to analyze. Experts looked for ways to make small, cheap improvements to the community that would help control water flow, such as installing gutters and digging ditches.

A landslide destroyed homes and covered part of a major highway in Colombia in 2016. The slide killed several people. Several more were reported missing.

CHAPTER 5

Conclusion

Landslides and avalanches take place all over the world. Although these natural events happen quite often, some are very small and may even go unnoticed. Others cause large amounts of destruction and deaths. However, people can take action to reduce the risk and limit the damage of slide events. Educating people about the hazards in their communities is one of the best ways to help them stay safe. Showing people how to identify slide hazards and warning signs and explaining how human activities trigger slides helps reduce the effects of natural disasters.

Scientists are always looking for new ways to predict when earth movements might take place so they can help people make better use of the land and give them advance notice of potential hazards. The more warning they can give locals to prepare or evacuate the area, the better. Thanks to new technologies, scientists know more today about landslide and avalanche processes than ever before. However, scientists also study the past to look for clues about the causes and effects of slide events.

We cannot always stop natural events from taking place, but we can try to reduce risk and their impact on people. It is important to alert people to the dangers they face in high-risk areas so they can make better choices about where they live and how they use the land. Knowing how to prepare for a slide event and what to do before, during, and after a landslide or avalanche helps saves lives.

A 1970 earthquake caused the largest ever avalanche at Mount Huascaran in Peru. The north side of the mountain collapsed, killing more than 20,000 people.

The largest landslide on Earth occurred at the eruption of the Mount St. Helens volcano in the Cascade Mountain Range, Washington, in 1980.

Ask Yourself This

Based on the information in this book, what are some of the things humans have learned from devastating landslides and avalanches and their aftereffects?

1. Why is it important for scientists to study earth movements? What tools and technology do they use?

2. Where do landslides and avalanches take place? Why are they more likely to occur in some places over others?

3. What can scientists learn by studying the locations of past landslides and avalanches?

4. Pretend you live on a steep slope, and the rainy season is about to begin. How can you find out if there are any slide hazards in your community? Are there any signs that might tell you if a landslide or avalanche is going to happen? What can you do to prepare for or help prevent a slide event?

Bibliography

Introduction

Birkeland, Karl W. "Avalanches." *Encyclopedia Britannica*. www.britannica.com/science/avalanche

Bonikosky, Laura Neilson. "Frank Slide: Canada's Deadliest Rockslide." *The Canadian Encyclopedia*, March 4, 2015. https://bit.ly/2tbZkQZ

"How many deaths result from landslides each year?" USGS. https://on.doi.gov/35XHiz3

Howard, Jenny. "Avalanches, explained." *National Geographic*, July 19, 2019. https://on.natgeo.com/388i7Lz

"Landslides and avalanches." Government of Canada. https://bit.ly/35Y4GMG

Live Science Staff. "What is a Landslide?" Live Science, December 24, 2012. https://bit.ly/35Ss8Ln

Sen Nag, Oishimaya. "Deadliest Landslides in Recorded History." WorldAtlas, April 25, 2017. https://bit.ly/2QWFWQY

"Snow Avalanches." National Snow & Ice Data Center, January 10, 2020. https://bit.ly/2RhAiYB

"Soldiers parish in avalanche as World War I rages." History, December 11, 2019. https://bit.ly/2NurGNf

"What is a landslide and what causes one?" USGS. https://on.doi.gov/2RqlgQq

"What was the biggest landslide in the world?" USGS. https://on.doi.gov/2NrMFQy

Chapter 1

"About the N.A.C." National Avalanche Center. https://bit.ly/3ad0RXv

"All About Snow." National Snow & Ice Data Center, January 10, 2020. https://nsidc.org/cryosphere/snow

Byrne, Kevin. "What is a lahar?" AccuWeather. July 10, 2019. https://bit.ly/2NtEPpv

Cohen-Waeber J., Sitar N. and Roland Bürgmann. "GPS instrumentation and remote sensing study of slow moving landslides in the eastern San Francisco Bay hills, California, USA." https://bit.ly/2NsPjFS

"Conditions for a slab avalanche." Mountain Creek Academy. https://bit.ly/2QUf3Nm

Guzzetti, Fausto. "Forecasting natural hazards, performance of scientists, ethics, and the need for transparency." ResearchGate, April 2015. https://bit.ly/35RrJJ3

Highland, Lynn M. and Peter Bobrowsky. "The Landslide Handbook—A Guide to Understanding Landslides." USGS. https://on.doi.gov/2RofRZS

Horton, Jennifer and Mark Mancini. "How Landslides Work." HowStuffWorks. https://bit.ly/2Rl0L7C

"How dangerous are pyroclastic flows?" USGS. https://on.doi.gov/387fn15

Meng, Xingmin. "Landslide." *Encyclopedia Britannica*. https://bit.ly/387hLoz

Migiro, Geoffrey. "What Is the Difference Between A Landslide And An Avalanche?" World Atlas, September 24, 2018. https://bit.ly/30lCuT1

"Natural hazards and disaster risk reduction." World Meteorological Organization. https://bit.ly/2tbWk7b

Roden, Barbara. "Golden Country: What caused the Hope Slide?" *The Ashcroft-Cache Creek Journal*, November 13, 2018. https://bit.ly/3aeZilE

"Topples." British Geological Survey. https://bit.ly/2uPIfwp

Chapter 2

Bolt, Bruce A. "Earthquake." *Encyclopedia Britannica*, January 9, 2020. https://bit.ly/3ah91OK

Fritz, Angela. "How the harrowing Thomas Fire planted the seed for California's deadly mudslides." *The Washington Post*, January 10, 2018. https://wapo.st/2FSkSVl

"How do landslides cause tsunamis?" USGS. https://on.doi.gov/35Zcdeis

"How Do Volcanoes Erupt?" USGS. https://on.doi.gov/2FOP43A

"India floods: More than 5,700 people 'presumed dead'." BBC, July 15, 2013. https://bbc.in/3afJrcA

Naranjo, Laura. "Connecting rainfall and landslides." Earthdata, Janaury 6, 2020. https://bit.ly/2Tpwgjl

Oskow, Noah. "Cataclysm: The 1888 Eruption of Fukushima's Mount Bandai." Unseen Japan, June 25, 2019. https://bit.ly/36Y5seb

Parker, Laura. "Will Everest's Climbing Circus Slow Down After Disasters?" *National Geographic*, May 13, 2015. https://on.natgeo.com/2NtkNvs

Riley, C.M. "Debris Avalanches, Landslides, and Tsunamis." https://bit.ly/3ab3FEs

The Editors of Encyclopedia Britannica. "Kilauea." *Encyclopedia Britannica*, March 24, 2018. https://bit.ly/2QVmaFb

The Editors of Encyclopedia Britannica. "Venezuela mud slides on 1999." *Encyclopedia Britannica*, December 15, 2008. www.britannica.com/event/Venezuela-mud-slides-of-1999

"Tsunamis." *National Geographic.* https://on.natgeo.com/387gRZd

"Why do earthquakes trigger catastrophic landslides?" *Encyclopedia of the Environment*, April 24, 2019. https://bit.ly/2QZqwvn

Chapter 3

Guzzetti, Fausto, Alessandro Cesare Mondini, Mauro Cardinali, Federica Fiorucci, Michele Santangelo, and Kang-Tsung Chang. "Landslide inventory maps: New tools for an old problem." ScienceDirect, 2012. https://bit.ly/35UE7It

Handwerger, Alexander L., Mong-Han Huang, Eric Jameson Fielding, Adam M. Booth and Roland Bürgmann. "A shift from drought to extreme rainfall drives a stable landslide to catastrophic failure." Scientific Reports, February 7, 2019. https://go.nature.com/2FUKXDf

"Hazard Mitigation." Alaska Department of Military and Veteran Affairs. ready.alaska.gov/Plans/mitigation

"Landslide Hazards." USGS. https://on.doi.gov/38fPiNz

"Landslide, Mud/Debris Flow, and Rockfall." Planning for Hazards: Land Use Solutions for Colorado. https://bit.ly/2TrCQWP

"Landslide prediction tool uses fibre-optic sensors." CBC News, September 29, 2014. https://bit.ly/2tmspsH

Martins, Daniel. "On Camera: Avalanche rolls through Italian village." The Weather Network, November 17, 2019. https://yhoo.it/2FT7e4i

"Overview of Hazards and Risk Assessments." USGS. https://on.doi.gov/2RlnMXP

"Pakistan Floods and Landslides – Mar 2016." ReliefWeb. https://bit.ly/30yC0sV

Palmer, Jane. "Slumgullion: Colorado's natural 'lab' offers insights into landslides worldwide." *Earth Magazine*, April 27, 2018. https://bit.ly/3adq0RN

"Prolonged closure of Karakoram Highway disrupts life in Gilgit-Baltistan." *Pamir Times*, April 12, 2016. https://bit.ly/377qG9D

Rasmussen, Carol. "Drought then Deluge Turned a Stable Landslide into Disaster." NASA Earth Observatory, May 27, 2017. https://go.nasa.gov/378GdFO

Schuster, Robert L. and Lynn M. Highland. "Impact of Landslides and Innovative Landslide-Mitigation Measures on the Natural Environment." USGS. https://on.doi.gov/2RoCQ74

Vienna University of Technology. "Rock Avalanches And Landslides: Modeling When The Mountain Slides Down Into The Valley." ScienceDaily, November 21, 2008. https://bit.ly/2uTyTzV

"Why study landslides?" USGS. https://on.doi.gov/30lCQZR

Wieczorek, Gerald F. and James B. Snyder. "Monitoring Slope Movements." National Park Service, January 19, 2018. https://bit.ly/2NoyM5Nv

Chapter 4

Erickson, Amanda. "Avalanches are becoming more common, thanks to climate change, researchers say." *The Washington Post*, January 19, 2017. https://wapo.st/2sqxVdq

Horton, Jennifer and Mark Mancini. "How Landslides Work." HowStuffWorks, June 6, 2018. https://bit.ly/2QWH1s0

Howard, Jenny. "Avalanche safety tips and preparation." *National Geographic*, August 5, 2019. https://on.natgeo.com/2Nut6Y5

"Landslide Preparedness." USGS. https://on.doi.gov/385hPVG

"Landslides & Debris Flow." Ready. https://bit.ly/2TmGZey

"Landslides: Before, During & After." Canadian Red Cross. https://bit.ly/3a9vRYm

"Online Avalanche Tutorial." Avalanche Canada. https://bit.ly/38kskoz

Smith, Harry, Soledad Garcia-Ferrari, Gabriela M. Medero and Helena Rivera. "Commentary/Hillside communities learn to mitigate landslide risks." RE.THINK, March 22, 2018. https://bit.ly/2FW96ck

Vinge, Vernor. "Ka-BOOM! 3 Types of Remote Avalanche Control in BC." TranBC, February 20, 2018. https://bit.ly/2u1Ush8

"What are Floods and Landslides?" RECARE-Hub. https://bit.ly/2RhA9Ex

"What is climate change." David Suzuki Foundation, October 5, 2017. https://bit.ly/2Tpx5bV

"What is the difference between a landslide advisory, a landslide watch, and a landslide warning?" USGS. https://on.doi.gov/2TsxwCk

Yle News. "Skiers who died in Arctic Norway last winter triggered avalanche: report." Eye on the Arctic, July 5, 2019. https://bit.ly/378Gx7u

Learning More

Books

Claybourne, Anna. *100 Most Destructive Natural Disasters Ever*. Scholastic Reference, 2014.

Gilbert, Sara. *Landslides*. Creative Paperbacks, 2018.

Johnson, Terry Lynn. *Avalanche!*. HMH Books for Young Readers, 2018.

Klimo, Kate. *Dog Diaries #3: Barry*. Random House Books for Young Readers, 2013.

Websites

Learn how to prepare for a landslide.
www.ready.gov/landslides-debris-flow

Find out what to do in an avalanche.
www.travelandleisure.com/travel-news/what-to-do-in-an-avalanche

Find out more about the work of the USGS to study landslides.
www.usgs.gov/natural-hazards/landslide-hazards

Watch a video of an avalanche in action.
www.nationalgeographic.org/encyclopedia/avalanche

Glossary

collapsed Fell down or caved in

deforestation The clearing away of trees and forests

deformations Changes in an object's shape or size due to a force being applied to the object

degradation The process of making something worse than it once was

droughts Long periods of dryness with no rain that leads to a lack of water

erosion The natural wearing away of Earth's surface over time due to natural factors, such as wind and water

evacuated Moved people from a dangerous place to a safer place

faults Cracks between Earth's crust that are found at the border between tectonic plates

fertile Capable of producing vegetation or crops

forces Actions that act on an object to change its state, such as set it in motion

friction A force that resists the motion of one object moving over another object

geological Related to Earth's structure and the substances that it is made from

Global Positioning System (GPS) A navigation system that can be used to find the exact location and speed of something any time of day and in all types of weather

gravity A force that pulls objects toward one another

hazard Something that can cause harm or is dangerous

infrastructure Basic systems, such as roads, buildings, schools, and utilities, that are part of a community

irrigation A system used to supply water to an area

magma Molten rock

magnitude Size or extent of something

momentum The force that keeps an object in motion

monsoon A season of strong winds and heavy rain in the Indian Ocean and Southeast Asia

optical fiber A thin plastic or glass fiber that allows light to pass through without much loss in its strength

permafrost Ground that remains permanently frozen

precipitation Rain, snow, sleet, and mist that fall to the ground from clouds

real-time Data that is sent at the same time as the event takes place

satellite An object sent into space to collect data as it orbits Earth

seismic activity The types, size, and frequency of earthquakes in an area

simulate To produce a computer model of something

tectonic plates Massive slabs of rock that make up Earth's surface and float over the mantle

tremors Slight earthquakes

tsunamis Massive waves that happen when a large amount of water is displaced from the ocean or a lake

utilities Services provided to the public, such as power, water, and sewers

vent An opening where air, gas, and liquid can escape

vibrations Fast, continuous, shaking motions

volume The amount of space something takes up

weathering The breaking down of minerals, rocks, soil, and other earth materials due to natural factors, such as wind and water

Index

About the Author

Heather C. Hudak has written hundreds of books for children on topics ranging from ancient history to modern technology. When Heather is not writing, she enjoys traveling the world and camping in the mountains with her husband and their rescue pets. She lives in a mountainous region where landslides often block major roadways throughout the year. Heather has also visited the Frank Slide site many times.